AF886681

AN OVERVIEW OF THE POST MINING MINERALS FOUND AT BISBEE, ARIZONA

POST-MINING IRON HYDROXIDES AND MELANTERITE COATING MINE TIMBER ON THE 2833 LEVEL OF THE CAMPBELL MINE

RICHARD GRAEME III–RICHARD GRAEME IV–DOUGLAS GRAEME

COVER PHOTOGRAPH TAKEN IN 1973 OF MINE TIMBER AND PIPE COATED BY POST-MINING IRON HYDROXIDE STALACTITES, PARTIALLY ALTERD TO GOETHITE, WITH MINOR TAN-YELLOWISH MELANTERITE ON THE 2833 LEVEL OF THE CAMPBELL MINE.

FORWARD:

As the name suggest, this is an overview of those minerals, which formed after mining had opened up new, previously buried areas. When mining advanced, it brought with it near-surface atmospheric conditions to the newly-opened areas, bringing oxygen, humidity and, importantly, certain bacteria into contact with the previously buried minerals, thereby allowing the process of oxidation to take place beginning with pyrite (FeS_2). Pyrite was the most common hypogene (primary) sulfide mineral deposited at Bisbee and is thermodynamically unstable and oxidizes, often quickly, as documented by the voluminous literature on the subject.

Figure 1: Typical massive sulfide ore - 2433 level, Campbell Mine.

Pyrite (brassy/yellow) is the dominate species present as abundant scattered grains and masses, along with chalcopyrite (lower right), a bornite/chalcocite/minor covellite mix (gray) and white quartz.

A variety of uncommon species consisting of largely copper/silver/gold/ sulfur coupled with tin/tungsten/antimony/vanadium/arsenic/bismuth and other elements are also present as accessory minerals in tiny, anhedral grains.

The process of post-mining mineral development is caused and often accelerated by the presence of oxygen and the water in the newly introduced atmosphere, while the bacteria, *Thiobacillus ferrooxidans* and *Thiobacillus sp* are usually present and involved. If conditions are ideal, this oxidation can proceed at a very rapid pace.

As noted, pyrite was always the dominant mineral in the hypogene sulfide deposits at Bisbee; both in the porphyries of the open pits and more so in the many, scattered limestone-hosted replacement deposits mined underground. Bryant and Metz, (1966) estimated that more than 500 million tons of pyrite had been deposited in the rock units hosting and surrounding the copper deposits at Bisbee.

It occurred as huge, compact replacement masses in the limestones as well as small areas of granular material. A wide variety of other sulfide minerals were deposited as well during the multiple phases of mineralization (Graeme, 1993). Large amounts of iron/copper/lead/zinc/sulfur bearing minerals replaced the hosting limestones, while iron/copper/sulfur minerals were deposited in the rock units comprising the Sacramento Stock complex.

The vast majority of pyrite found during mining was left in place underground, because it was not ore, but rather waste, as it contained little to no copper mineralization. To be sure, some pyrite was intermixed with the ore minerals and mined, but this was a relatively minor amount when compared to the total volume of pyrite present in a typical hypogene replacement deposit at Bisbee. In the Sacramento and Lavender open pit mines, the ores were very different and occurred as a very thin, supergene (secondary) coating of chalcocite and similar secondary copper sulfides on pyrite. Thus, unlike the underground, substantial pyrite was mined with the copper sulfides.

As pyrite oxidation is an exothermal process, the exposed pyrite generated heat during oxidation, often substantial heat. Mining through the transition zone, where oxidation was naturally occurring from oxides to sulfides, it was always hot, with air temperatures frequently reaching $90^0 +$ F. Underground, in areas with granular or sand-like pyrite, the large surface area presented by the many particle faces promoted rapid oxidation, generating sufficient hear to ignite the supporting timbers or even self-combust by friction generated heat, if ground movement occurred (Mitke, 1920).

Pyrite fed mine fires were a constant problem at Bisbee from about 1900 on to the end of mining. Water pumped into the burning areas, exited with a low pH and highly charged with iron and copper and was captured for copper recovery. Later, interesting post-mining minerals formed where these waters ponded or seeped through surrounding rock.

There can be little doubt that many, if not most, of the minerals species, now only recognized as post-mining, had formed during natural oxidation process before mining began. Further many of these species were surely encountered during mining in the transitional zone between the oxides and sulfides.

However, given the general lack-luster appearance of most such species and their typically high water solubility, they went unnoticed. And too, these often unstable mineral species had to compete spectacular azurite and malachite specimens, which were always desirable.

Figure 2: Mining fresh sulfide ores - 2833 level, Campbell Mine. Peter Kresan photo

Calcite, unless very colorful, and copper were similarly ignored until the late 1950s when other, more attractive minerals were less common in the mines. Very few copper or calcite specimens from early mining at Bisbee exist, as few were saved.

INTRODUCTION:

Post-mining mineral deposition occurs in several ways. At Bisbee, most such minerals are sulfates and deposited as the natural processes of sulfide oxidation continued during and after miners made their openings for either exploration or exploitation. Minerals other than sulfates formed as post-mining deposits as well, but were relatively minor in nature. In some cases, the non-sulfate, post-mining minerals were also a direct result of the sulfide oxidation, such as causing elemental copper to replace iron/steel pipe and rail. In reactive environments, such as limestone, typically carbonates formed. Natural ground water channels interrupted by mining caused aragonite and calcite to be deposited in the mine workings, often in substantial volume.

Fire related minerals were few, as the underground fire areas were usually flooded with water or a slurry of mill tailings before being reopened. This would dissolve any of the hydrated sulfates.

A list of those species recognized as forming under post-mining conditions is presented below:

alunite	boothite	cuprocopiapite	hydrozincite	plumbojarosite
alunogen	botryogen	epsomite	ilsemannite	rhomboclase
anglesite	brochantite	ferricopiapite	jokokuite	römerite
anhydrite	calcite	fibroferrite	lime	rozenite
anthonyite	chalcanthite	goethite	malachite	siderite
antlerite	chalcocite	goslarite	melanterite	siderotil
aragonite	chrysocolla	gypsum	metavoltine	sulfur
aurichalcite	copiapite	halloysite	paratacamite	szomolnokite
azurite	copper	halotrichite	paracoquimbite	voltaite
basaluminite	coquimbite	hexahydrite	pickeringite	zincobotryogen
bianchite	cuprite	hydrobasaluminite	ransomite	zincocopiapite
bilinite				

Table 1: Post-mining mineral species recognized at Bisbee

More detail on the post-mining mineral occurrences at Bisbee can be found in Anthony, et al., (1995), Graeme (1981, 1993), (Graeme, et al., 2015), Merwin and Posnjak (1937) and Mitchel (1921). Unless otherwise noted, all photographs are by the authors and the specimens are from the Graeme collection.

SULFATES:

The vast majority of the recognized post-mining species are water-soluble, hydrated sulfates. Post-mining sulfates were widely distributed throughout the mines and locally abundant. These sulfates develop as a result of the decomposition of the sulfides, hypogene or supergene, in the moist, now oxygenated mine environment. Pyrite is the first of the sulfides to be impacted by the newly introduced conditions. The decomposition of pyrite under damp conditions, generates small amounts of low pH, iron sulfate solutions that migrate along fractures and crystal boundaries to the mine openings. When present, other sulfides react with these iron sulfate solutions liberating other elements, particularly copper or zinc, albite more slowly.

Figure 3: In situ, massive pyrite decomposing and generating low pH solutions, which are forming iron hydroxides (brown) and have begun to mobilize copper - copper iron sulfates from other sulfide minerals. View-30 cm, 6 level Southwest Mine.

Evaporation of the solutions from the mine walls and floors typically forms multi-mineral crusts with the lower pH phases at the wall surface grading to higher pH phases at the outer part of the crust, reflecting evaporation and, consequently, changing of the pH, as discussed by Merwin and Posnjak (1937). Non-reactive wall rocks such as porphyry, massive sulfides or silicified units are important to the development of these crust growths. While these post-mining crusts often contained multiple species, it was not uncommon for one species to be decidedly more abundant, with many of the others little more than accessory species.

In some cases, large areas of non-reactive wall rock were covered by post-mining mineral crust. In moist, high pyrite areas, the authors have seen tens of square meters covered by post-mining sulfates.

Figure 4: A post-mining, multi-mineral crust continuing at least - copiapite $Fe^{2+}Fe^{3+}(SO4)_6(OH)_2 \cdot 20H_2O$: (yellow), melanterite (cuprian) $Fe^{2+}SO_4 \cdot 7H_2O$: (Green), kornelite $Fe^{3+}(SO_4) \cdot 7H_2O$: (very light purple). In situ, 7 level, Southwest Mine, view - 16 cm.

Figure 5: A post-mining, multi- mineral crust containing coquimbite, $Fe_2{}^{3+}(SO_4)_3 \cdot 9H_2O$: (purple), rhomboclase, $HFe^{3+}(SO_4)_2 \cdot 4H_2O$: (yellow-white), botryogen, $MgFe^{3+}(SO_4)_2(OH)_3$: (orange) 1600 level, Campbell Mine, view – 3.2 cm.

Figure 6: Post-mining crust of coquimbite, $Fe_2{}^{3+}(SO4)_3.9H2O$: (light purple), paracoquimbite, $Fe_2{}^{3+}(SO_4)_3 \cdot 9H_2O$:(light purple), copiapite, $Fe^{2+}Fe^{3+}(SO_4)_6(OH)_2 \cdot 20H_2O$: (yellow), melanterite $Fe^{2+}SO_4 \cdot 7H_2O$: (white), voltaite, $K_2Fe_5{}^{2+}Fe_4{}^{3+}(SO_4)_{12} \cdot 18H_2O$: (black). Lavender Pit Mine, view – 8 cm.

Single species development along mine openings was quite common as well. Typically, the areas that hosted a single or very few post-mining species were somewhat less moist and/or contained only modest amounts of sulfide minerals. Thus, the oxidation and the development of the post-mining mineral/minerals was both slower and more limited. Other factors such as a single sulfide species present could also contribute to the development of just one post mining species or one dominate species.

Figure 7: Melanterite, $Fe^{2+}SO_4{\cdot}7H_2O$, coating high-pyrite wall rock. 6 level, Southwest Mine.

Figure 8: In situ halotrichite, $Fe^{2+}Al_2(SO_4)_4{\cdot}22H_2O$:, 1600 level, Campbell mine:, View – 50 cm.

Figure 9: (left) White epsomite, $MgSO_4.7H_2O$:, effervescing from Martin limestone, 6 level, Southwest Mine: view – 1.35 meter

Figure 10: In situ, blue ransomite, $Cu^{2+}Fe_2^{3+}(SO_4)_4 \cdot 6H_2O$:, with minor white melanterite , 1800 level, Campbell mine: view - 35 cm.

Under conditions where more water is available, other types of growths develop with distinct crystals uncommon. Stalactites and less commonly stalagmite forms were common where dripping solutions were present. These forms are generally composed of a single mineral species that is dominant in the area and reflecting the source mineralogy; due to a single source for the solutions and a relatively consistent pH, as opposed to multi-mineral crust in moist areas.

However, it was not particularly uncommon to find copper, zinc, or iron-containing varieties of the dominant species locally intermixed with the more prevalent species. An example is the presence of pisanite (cuprian melanterite) stalactites was noted with similar stalactitic forms of melanterite and/or chalcanthite. (See Figure 12.)

Figure 11: Chalcanthite, $Cu^{2+}SO_4 \cdot 5H_2O$: as stalactites and stalagmites. In a mined out stope. 2300 level, Junction Mine.

Figure 12: Chalcanthite, $Cu^{2+}SO_4 \cdot 5H_2O$: (blue), with cuprian melanterite, $Fe^{2+}SO_4 \cdot 7H_2O$: (near - black green) on copiapite, $Fe^{2+}Fe^{3+}(SO_4)_6(OH)_2 \cdot 20H_2O$: (yellow) coated mine timber. 2833 level, Campbell Mine, vertical view - 61 cm.

Figure 13: Goslarite, $ZnSO_4 \cdot 7H_2O$:, as stalactites on mine timber. 1500 level, Junction Mine. Vertical view - 1.25 meters.

Broken sulfides scraps left underground in damp, non-reactive environments, often developed sulfate coatings. These ranged from thin veneers to layers a centimeter thick. Lumps of massive pyrite frequently completely decrepitated, while relatively pure massive bornite, chalcopyrite and chalcocite had only a surficial alteration to brochantite and/or antlerite, devilline or chalcanthite, respectively. Galena acquired a thin coat of anglesite and the typically iron-rich sphalerite formed goslarite and a few other sulfates, depending on the relative abundance of pyrite. However, even modest amounts of intermixed limestone with the broken sulfides either inhibited or prevented sulfate crust formation altogether through buffering the acidic formation.

Figure 14: Chalcanthite, $Cu^{2+}SO_4 \cdot 5H_2O$: (blue) on chalcocite and brochantite, $Cu_4^{2+}(SO_4)(OH)_6$: (green) on bornite as alteration coatings on broken ore left behind. In situ. 6 level. Southwest Mine. view – 1.20 meters.

Figure 15: (left) ponded, low pH water with iron hydroxides forming on the surface and edges. Tunnel level, Higgins Mine.

Figure 16: (right) ponded low pH waters depositing iron sulfates due to evaporation. 2833 level, Campbell Mine.

For scale; the mine workings are approximately two meters wide.

Where acidic solutions formed stagnant ponds, it was common to find crystals, often large, of several of the sulfates intermixed. Chalcanthite, melanterite and large, colorless epsomite crystals were the most common species in the ponds as strikingly beautiful, but ephemeral specimens.

Figure 17: Collecting cuprian melanterite and epsomite crystals from ponded mine waters on the 2700 level of the Junction mine.

A change in the mine environment to warmer, dryer conditions often occurred as the available water drained from the rock, and continued sulfide decomposition, an exothermic process, warmed the air. Modifications in mine ventilation also altered the nature of the environment as well by changing the airflow and bringing in drier or increased volumes of air. These changes frequently caused dehydration of the original sulfates, with new species such as hexahydrite replacing epsomite and rozenite or siderotil replacing melanterite.

Figure 18: Römerite, $Fe^{2+}Fe_2^{3+}(SO_4)_4 \cdot 14H_2O$: growing on pyrite-rich porphyry in the Lavender Pit Mine, View-80 cm.

The exposed sulfides in remaining porphyries and the heavily pyritized Pinal schist of the Lavender Pit Mine develop an interesting, albeit typically short-lived, suite of post-mining minerals as well. The periodic rains filter through the broken wall rock and form spotty effervescent growths of usually iron sulfate species. However, bluish blossoms of ilsemannite can be seen on occasion, betraying the highly localized and uncommon presence of molybdenum. For the most part, these sulfate growths only last until the next rainfall, when they are dissolved or the more typical dryness of the Arizona air returns, which causes their dehydration and decrepitating.

Very few examples of the hydrate sulfates from Bisbee have been preserved, as most quickly dehydrate on exposure to the dry, surface atmosphere and either alter to other, less hydrated species or completely decrepitate due to the loss of water from the mineral structure. Many were the miners or mine visitors who collected brilliant blue samples of the abundant chalcanthite or cuprian melanterite, only to have it turn into an uninspiring white, chalklike mass within a few days in the dry surface atmosphere as it lost its water of hydration and altered to another mineral, usually white or yellow/white.

This tendency for the rapid dehydration of some of the sulfates has complicated their study. When collecting these minerals, one must go prepared to preserve, as much as possible, the humid conditions before removal from the location of deposition by using airtight containers. Then, storage, once removed from the mine, must be considered. Even with the best of planning and efforts, most of these species will still dehydrate with time. Indeed, few truly remain unaltered.

Fortunately, a few sulfate species do survive. Excellent examples of römerite, voltaite, copiopite, coquimbite, paracoquimbite, botryogen, szomolnokite, and zincobotryogen are in several collections. Post-mining antlerite, brochantite, devilline and langite are also well represented as interesting, but usually uninspiring specimens.

Many hundreds of post-mining gypsum specimens were collected at Bisbee and are frequently seen in collections. Gypsum, as the variety selenite, was common in stagnate ponded, low to moderate pH mine waters as well, but never in association with an abundance of other sulfate species, as the necessary calcium carbonate partially neutralized the solutions. Selenite often occurred as a growth on limestone walls of flooded drifts or on the bottoms of the ponds, but always on or very near limestone.

Figure 19: Ponded acidic waters to 25 cm. deep, in a limestone crosscut, 6 level, Southwest Mine, vertical view - 2.30 meters.

Abundant yellow/white, post-mining selenite had/was forming at the limestone acid water interface as thick masses of crystals, often exceeding seven centimeters in length.

Figure 20: Gypsum $CaSO_4{\cdot}2H_2O$:, as selenite on post-mining copper, 2200 level, Campbell Mine, specimen - 7cm.

This specimen formed in a very small pond of low pH-high copper solutions in a limestone crosscut bottom. Little to no iron was present in the solutions, something not too common.

One other post-mining sulfate worth noting was a single occurrence of post-mining alunite, which was an actively oozing from a fracture on the 1300 level station of the Cole Mine in the 1960s. It was light blue in color and frequently "harvested" by the miners as it grew in size. However, once on the surface for a few days, it would craze upon dehydration, becoming quite fragile and much lighter in color.

IRON HYDROXIDES:

Figure 21: Acidic mine water depositing iron hydroxides in abandoned working, note abundant white to brown melanterite on ceiling and walls, 6 level, Southwest Mine, horizontal view – 2.8 meters.

Figure 22: Acidic mine water formed by the oxidation of pyrite. Mud-like iron hydroxides have precipitated for the water and partially filled the mine opening. Tan colored melanterite locally coats the mine walls. 7 level, Southwest Mine, horizontal view – 2.2 meters.

Pyrite oxidation in wet areas usually generated copious amounts of various iron hydroxides. The precipitation of the gelatinous iron hydroxides was a problem caused by these low pH solutions. If left unattended, mine workings would fill to a depth of a meter or more with the troublesome slime. It is this iron laden, low pH water that is much a part of the environmental problems associated with pyrite degradation.

Ponded mine waters in sulfide areas collected behind rock falls or mine timber failures in abandoned areas. The dammed up solutions behind these obstructions often filled with a thick layer of yellow mud, consisting of iron hydroxides, which had precipitated out of the high-iron solutions. Sometimes the mud would be waist-deep. If undisturbed, goethite rafts formed in still areas and eventually sank to the pond bottoms.

While some of these hydroxides were deposited as little more than a slimy mud in the workings, others often formed numerous long stalactites and massive stalagmites. Undoubtedly, some of the same iron compounds were deposited in both

Figure 23: Iron hydroxide stalactites and stalagmites deposited by the acidic water that have partially filled the mine opening. 7 level, Southwest Mine, vertical view – 2.3 meters.

Figure 24: Acidic mine water with stalagmite forms and minor melanterite. 6 level, Southwest Mine, view – 38 cm.

cases, but no effort was ever made on the part of the authors to determine either the chemistry or mineralogy of any, but rather to simply document the appearance during deposition and in some instances, note the ultimate apparent end minerals, which formed with dehydration.

In areas with substantial dripping water, which had passed through the oxidizing sulfides, impressive clusters of yellow to yellow/brown to dark brown stalactites would form. In several instances, the stalactite/stalagmite development was such that it nearly closed the mine opening.

Figure 25: Goethite coating "limonite" stalactites that formed similar to those in Figure 23. Specimen – 11 cm. Southwest Mine. This specimen is not of post-mining origin.

For the most part, the stalactitic growths were delicate; not soft, but rather hollow and/or composed of successive layers of material that were poorly cemented together. Often straits of solution were in between the concentric, crustlike layers.

On drying over an extended period of time, the iron hydroxides became an impure mixture of several iron oxides, but were predominantly goethite and lepidocrocite. In any event, the collective term "limonite" best describes the appearance of the ultimate product.

CARBONATES:

Aragonite:

The most common of the post-mining carbonates was aragonite. During the course of mining through the barren limestone host rock, it was common to intercept cracks, faults and solution channels that served as conduits for the usually small amounts of water within the limestone. Most often, the water in these openings was clean and metal free, but high in dissolved calcium and CO_2. As the water entered the mine openings, it began to liberate the CO_2 and deposit calcium carbonate, usually as aragonite, though calcite was deposited in some areas. The reason for the apparent significant preferential formation of aragonite over calcite is unclear, but may well be related to the relatively high kinetic energy associated with the deposition – cascading or fast flowing or long vertical drop of the depositing waters.

Figure 26: A one-gallon bucket coated by post-mining aragonite. 6 level. Southwest Mine.

In limestone areas, post-mining aragonite was a common occurrence in most of the mines. Aragonite stalactites of several centimeters in length, often coated with small, acicular aragonite crystals, were locally abundant where the groundwater flow was modest. In mine openings that served as main ventilation ways, these stalactitelike formations would grow directed into the strong airflow almost parallel to the mine back (ceiling), but without the acicular crystal growth on their exterior. With moderate ground water available, the walls and back could be covered with parallel aragonite growth reaching ten centimeters and all pointed into the wind, where evaporation took place the most rapidly.

The apparent rapid rate of deposition of aragonite was noted is several places, but perhaps the most interesting in our view, is a single, one gallon, galvanized steel bucket left by lessees in early 1944 on the 6th level of the Southwest Mine. It had been placed under a steady drip of clean water to use for drinking, as suggested by the remains of a wooden water keg close by. It would have taken several hours for the bucket to fill, yet when first found some 30 years later, a full three centimeters of semi-pearlescent aragonite had been deposited where the water flowed from the vessel. Today, more than five centimeters of aragonite have encrusted the lip of one side and the whole of the bucket is coated, representing some two centimeters of additional deposition in less than 40 years.

Thick, massive post-mining aragonite was abundant in several places where large volumes of mine water were transferred from level to level down 100-foot raises. Ladders and other timber were often coated with 5-10 cm of aragonite by the forceful, cascading waters. The most memorable are a raise between the 300 and 400 levels of the Czar Mine that channels water from under Queen Hill. These clean waters were kept separate from other mine water and collected for use on the surface in boilers, showers, etc., during the operation of the mine.

Near pure-white aragonite to 10 centimeters thick coated all of the timbers, in spite of the force of the hundreds of liters per minute falling down the raise. The receiving crosscut was flooded, as the 400 level station had collapsed sometime in the early 1950s, so the timber below the water level was somewhat less heavily coated by aragonite. As the rock damming up this significant flow was very course, the water passed through it easily, precluding the formation of aragonite/calcite rafts, so commonly seen covering still mine waters. Fortunately, good, natural ventilation kept the concentrations of liberated carbon dioxide low, allowing entry, as the water nearly filled the crosscut, forcing one to swim in the frigid water.

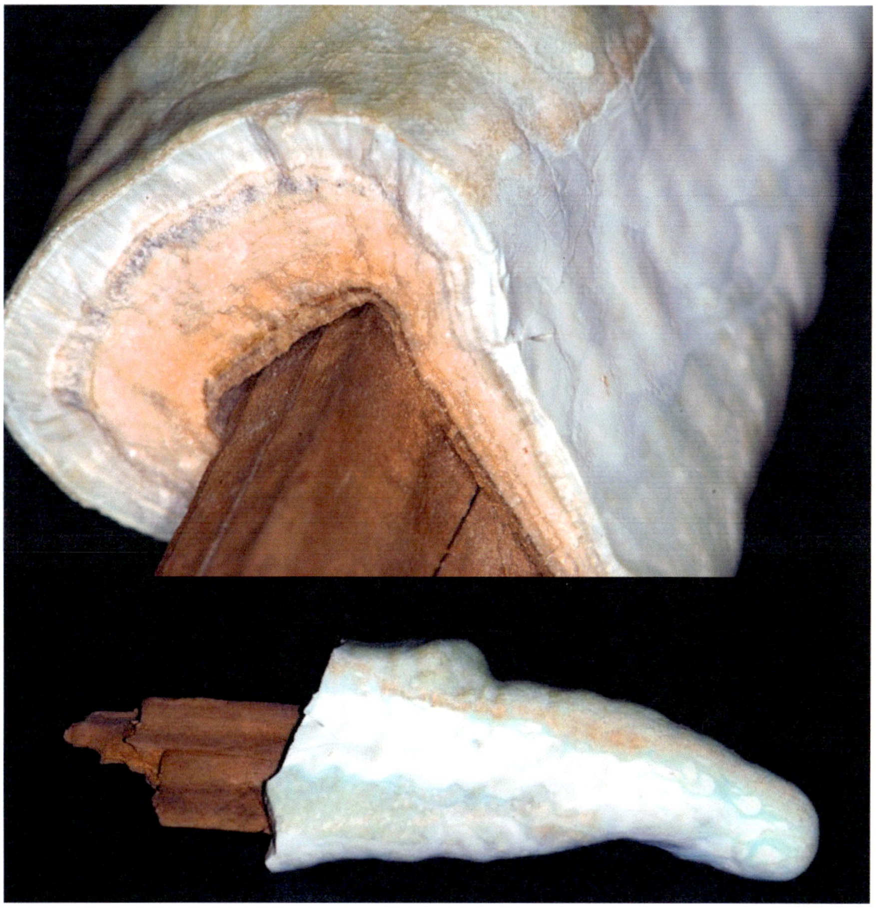

Figure 27: Two views of an 8 cm. aragonite stalactite-like form that grew over broken mine timber in about 25 years. The color variation in the top view, reflects changing water quality from turbid to clear. 2300 level, Junction Mine.

A second, but very similar site was on the 2300 level of the Junction Mine where the waters from the Denn Mine poured down a timbered raise from the level above on their way to the 2200 level pump station. For the most part, this water came from the wide Dividend Fault zone, cut when sinking the Denn Shaft in the 1930s. Over time, the water quality appears to have change, no surprise as substantial mining was taking place. The change in water quality is reflected in the variable coloration of the deposited aragonite. As mining passed well below the 2000 level, the flow decreased and some coloration due to copper imparted. By the early 1970s, just a few hundred liters per minute were flowing down this raise, a fraction of what company records noted some 30 – 40 years before.

Figure 28: Post-mining aragonite "cave pearls" to 2 cm, 2300 level. Junction Mine.

The diminished flow allowed for the formation of colorful blue/green rafts along the still margins of the ponded water. Similarly colored cave pearls to two centimeters in size formed where a small flow of water fell from a displaced mine timber into a small, otherwise still pool.

As exploration east of the Campbell Mine progressed in the early 1940s, substantial amounts of water were hit, as expected. While every reasonable precaution had been taken, the Campbell Mine was partially flooded in 1941 by a flow from 12 crosscut (XC) on the 2700 level, estimated at more than 200,000 liters per minute (Mills, 1958).

Figure 29: Water door on 12 XC, 2700 level, Campbell Mine in 1962. Note the abundant post-mining botryoidal aragonite between the mine rails. An example of this material is shown in Figure 28, below.

After months of pumping, access was regained, but substantial water still flowed from a series of subparallel faults in the mine wall. By 1960, thick, travertinelike aragonite as much as three centimeters thick coated the crosscut floor and the water was still flowing 25 centimeters deep across the whole of the two meter wide crosscut floor. A match lit a meter above the flowing water would quickly be extinguished lowered a bit by the high CO_2 concentrations due to the rapid degassing of the ground water. This degassing, of course, caused the relatively rapid deposition of aragonite.

Specimens of the tan to yellowish to gray,

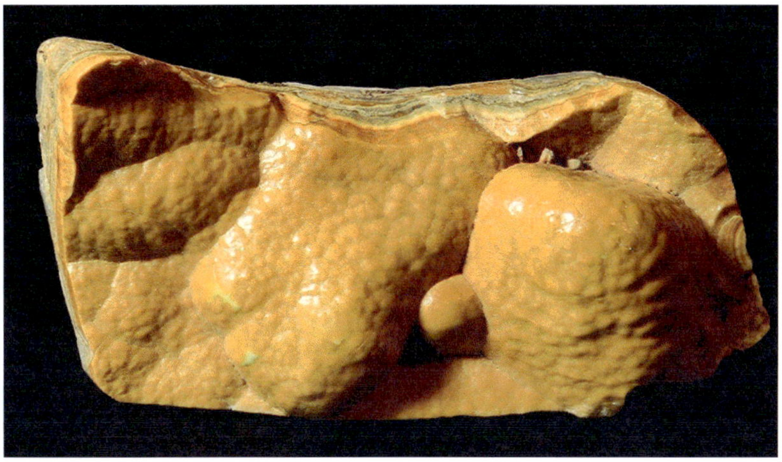

Figure 30: Post-mining aragonite from the floor of 12 XC, 2700 level, Campbell Mine. Specimen – 7 cm.

flowstonelike aragonite collected from the fast-flowing water often contain angular fragments of fresh limestone or sulfides as well as anything else thrown aside, such as wood, nails and broken light bulb pieces. Even pieces of forgotten native copper, which were placed in the water for cleaning (a common practices), have been covered by this aragonite.

On the 2200 level of the Campbell Mine, an unusual blue colored post-mining aragonite/calcite mixture formed in small amounts on and under a leaky wooden flume. The flume was constructed to capture water pouring from a two-inch diamond drill hole and carry it over a gobbed stope to a water ditch. This aragonite clearly shows ripple-like features and is occasionally stalactite when it formed on the bottom of the flume. The reason for its striking blue color is unknown, but it is

Figure 31: Aragonite/calcite – Front and back views of a 13 cm. post-mining aragonite and calcite mixture. This formed on a leaky, timber flume used to conduct water over broken ground. The pattern of the wood is clearly visible on the back side. 2200 level, Campbell Mine.

The intense blue color of this material caused it to be confused with azurite, but it is largely aragonite more or less altered to calcite.

reasonable to assume a high degree of copper substituting for calcium in the crystal lattice. More than a few have thought this material was azurite.

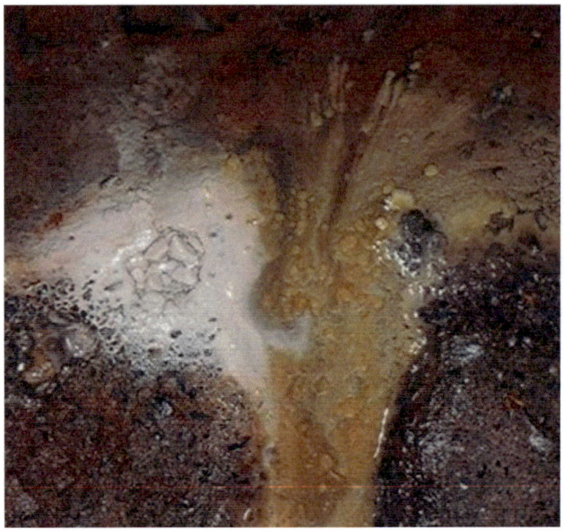

Figure 32: Aragonite "Bird's nest," forming adjacent to a pool of water. 6 level, Southwest Mine, view - 25 cm.

From a mineral collector's point of view, the post-mining aragonite forms the miners call "bird's nest" are a much coveted prize. These interesting forms developed in inactive, limestone areas of all of the mines, in still-air environments and from slow dripping solutions. These were flat deposits, cementing small areas of the loose rock of the mine floor, with a depression developed where the constant water drops hit. Small rock fragments within the depression would be coated with successive layers of aragonite, resulting in rounded, growths, much like cave pearls that resembled small eggs. Thus the local name.

These interesting forms were usually white, but bluish, copper-tinted, and red to brown; iron-colored examples were not uncommon. While "bird's nest" were not truly rare in the mines, successfully collecting one was a real challenge. Thus, few good examples exist. It took a number of years, often decades, for the slow deposition of aragonite from a single dripping source, to become thick enough to withstand the force needed to remove one; even though they formed on broken rock or compacted fill on the mine floor.

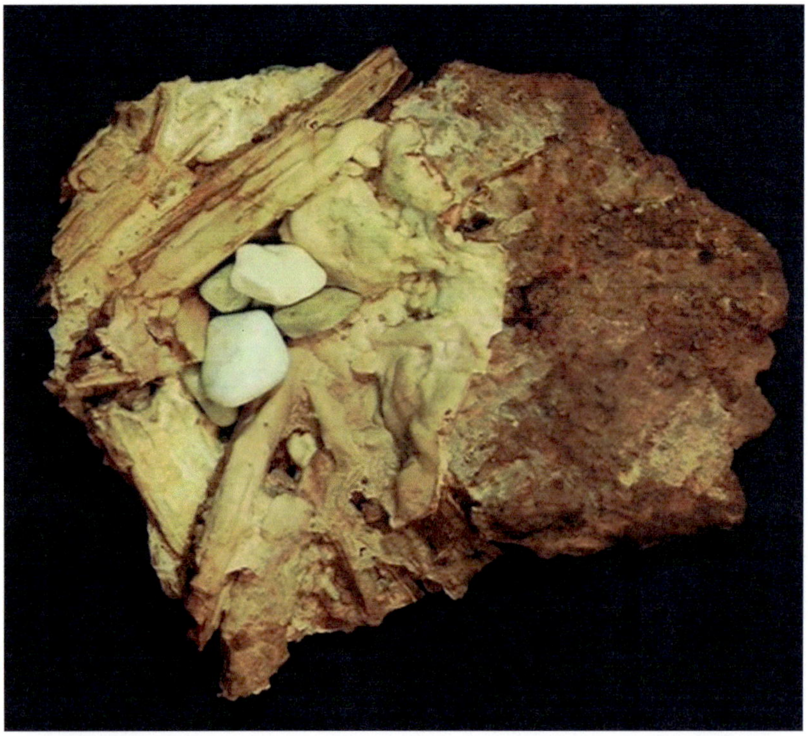

Figure 33: Aragonite "Bird's nest," with splinters of mine timber, 2200 level, Campbell Mine, specimen - 9 cm.

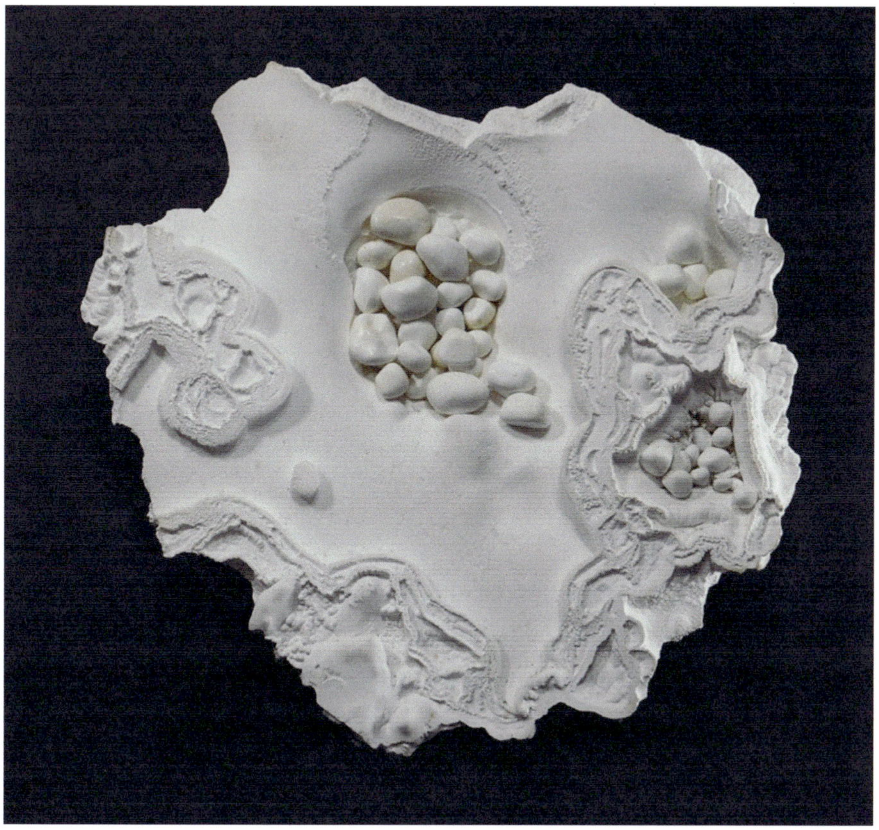

Figure 34: An uncommonly fine aragonite "Bird's nest," 2200 level, Campbell Mine, specimen - 27 cm

Miners were known to find a developing "bird's nest" and hid it by placing old mine timber to protect it from view and/or prevent someone from stepping on the delicate treasure. They also often began the task of preparing for removal by driving large nails around the area to develop cracks and would use strips of canvas to control the splash area into the area of interest. More than one miner would add a small bit of chalcanthite on occasion to enhance the color. However, no matter the care and planning taken, it was usually luck that allowed a successful recovery. Over our many years of viewing Bisbee minerals, we have seen fewer than ten such specimens that are of any real aesthetic interest.

Calcite:

Post-mining calcite was also widely distributed and common in the majority of the mines, typically forming small stalactites and thin flowstonelike ribbons on mine walls. Oddly, post-mining few crystals of calcite, other than microscopic size, are known from Bisbee. Very few post-mining calcite specimens from here were of interest to collectors.

Figure 35: (right) Typical post-mining calcite as small stalactites and flowstone on slightly altered Martin limestone, 6 level, Southwest Mine, view – 1.4 meters.

Figure 36: Post-mining calcite deposited by steadily seeping water along a wide fault zone. 2200 level, Junction Mine.

Most stalactites above aragonite "bird's nest" formations were calcite. Calcite also formed as thin "raft" forms on top of ponded clean water. Calcite was noted coating a long abandoned mine car under a small, but steady flow of water on the Tunnel level of the Higgins Mine, a most unusual occurrence. In Figures 38 and 39 are two photographs of this mine car taken a bit more than 40 years apart.

Figure 37: (left) A close-up view of the post-mining calcite in Figure 35, above on the 2200 level, Junction Mine. View – 30cm. Note the oriented growth direction. It was into the strong airflow caused by the mine ventilation, as a huge fan was less than 100 meters distant at the Junction Shaft.

Figures 38: (left top) Post-mining calcite growth on a mine car, Higgins Mine, Tunnel level in 1967. More than 1/2 inch of calcite had been deposited in the just over 20 years since this mine closed.

Figure 39: (left bottom) The same mine car some 40 + years after the first photo; illustrating the relatively rapid calcite deposition.

The calcium carbonate rich water is flowing from a small diameter diamond drill prospecting hole.

Figure 40: Colorless to white, post-mining calcite on post-mining aragonite, 6 level, Southwest Mine. Specimen – 9 cm.

Malachite:

Low pH, copper laden mine waters were a significant problem at Bisbee as early as 1890 when the mining of deeper ores exposed areas undergoing sulfide oxidation. Some of the water contained several tens of pounds of dissolved copper per thousand gallons of water. Any iron or steel tool or mine rail that was in this water for even a short time, was attacked with the iron going into solution and a useless copper form left in its place. Control of acidic mine waters, locally referred to as "copper water," was a critically important operating issue, even to the end of mining in 1975.

During mining early on, it was a common practice was to channel the acidic, metal-rich solutions through old, abandoned workings, including gobbed (backfilled) stopes, to eventually bring them to a pump station for removal to the surface for copper recovery, via iron precipitation. The gob in the early stopes was typically a mix of mine waste rock, much of it hand-cobbed limestone, sorted during mining. As early mining took place close to the Sacramento Stock Complex, hydrothermally altered limestone and porphyry waste were common in the gob as well.

Figure 41: Malachite as a replacement rind on recrystallized Escabrosa Limestone, from the Holbrook Mine, specimen - 13 cm.

It did not take long for the acidic waters to attack the limestone chunks in the gob, leaving a replacement veneer of usually malachite, $Cu_2^{2+}(CO_3)(OH)_2$, to several centimeters thick. In some cases, the typical process for forming malachite occurred with the development and transport copper carbonate solutions, typically gel-like, to open spaces for deposition. Such material is usually indistinguishable from most banded malachite. In gob exposed during the mining of the Lavender Pit mine, one of the authors found a piece of thick, banded malachite with a rusted track spike imbedded within, reflecting its post-mining origin.

Figure 42: Malachite as an impure replacement rind on Martin Limestone. Note the relic features of the limestone are preserved in the replacing malachite. Specimen – 7 cm.

Sufficient copper was deposited in a number of cases to make the gob ore grade (Bateman, et al. 1914). Douglas (1909) did not recognize the process and suggested that what was now ore had been placed as fill in the past. This also caused an erroneous belief among miners that "the old timers used malachite as a backfill," a local myth that persist to this day.

Several other copper minerals were deposited as well by the "copper water" passing through the gobbed stopes with copper and cuprite fairly common as well as minor azurite. Bateman, et al. (1914) noted that the acidic solutions altered the less reactive rock in the fill to several of the clay minerals such as kaolinite and halloysite.

In other environments, malachite often developed as a thin, often scaly, crust on limestone or calcite when copper-rich solutions flowed over them or in the residual broken rock adjacent to oxidizing sulfides in mined-out stopes. On rare occasion, stalactites of malachite formed on the edge of a broken rock face. In one instance, malachite was apparently being co-deposited with chrysocolla, as modest amounts of copper-laden water were comingled, with one part coming from a silica breccia and the other from fresh limestone. However, in general, post-mining malachite from Bisbee was seldom collected, as it was typically less than inspiring.

Azurite:

Post-mining azurite, $Cu_3^{2+}(CO_3)_2(OH)_2$, at Bisbee was rare. No doubt, this was a function of the higher CO_2 required for deposition than is typically found in normal mine atmospheres. Nonetheless, in a few instances it has been recognized as a surficial replacement or along fractures in limestone along with more abundant malachite. Subaerial deposits of azurite seem to mimic malachite in that they are quite gel-like when deposited.

In one extraordinary occurrence on the 3rd level of the Southwest Mine, where solutions emerging from a fracture had deposited botryoidal azurite over an area of more than a meter. While it was but a thin crust of three to four millimeters thickness, it was strikingly beautiful.

Other carbonate species that have been noted forming with azurite under post-mining conditions include aurichalcite. Post-mining aurichalcite $(Zn,Cu)_5(CO_3)_2(OH)_6$: was observed forming over

Figure 43: One of the authors on the 3 level, Southwest Mine where azurite being deposited in a gel-like flow from fractured limestone.

the period of just a few days along with azurite and malachite. The aurichalcite was formed from a thin jellylike mass running into an open space and drying in a 30-cm wide goethite seam bounded by limestone (Tenney, 1913).

Hydrozincite, $Zn_5(CO_3)(OH)_6$, has been noted as a localized post-mining flow from the alteration of massive sphalerite in the Copper Queen Mine. It was also reasonably abundant as an alteration product of the massive sphalerite used as waste backfill in a small stope on the "B" level of the Copper Queen Mine. At the time the Copper Queen Mine was operating (1880-1888), the limited zinc mineralization found was not of any economic importance, and thus they were discarded as waste.

Figure 44: One of the authors on the "B" level, Copper Queen Mine where ultraviolet lighting causes the massive sphalerite to fluoresce orange-red, calcite gives a red color while the post-mining hydrozincite responds to UV light with a blue-white fluorescence, and can be seen at center left.

Chrysocolla:

As noted, malachite, azurite and aurichalcite formed from gelatinous materials that typically oozed slowly and in small amounts from cracks as well as other small openings. Chrysocolla $(Cu^{2+},Al)_2H_2Si_2O_5(OH)_4 \cdot nH_2O$, formed in a similar manner. Indeed, post-mining chrysocolla was quite common, having been found in several mines as light blue stalactitic growths and as bright blue-green flows on opening walls. Without exception, this material crazes and becomes much paler in color due to desiccation when collected and taken to the surface

Figure 45: (left) Chrysocolla and malachite being deposited simultaneously.

Silica-rich solutions are coming from a silica breccia on the left, while carbonate-rich solutions are flowing from limestone on the right side.

6 Level, Southwest Mine, longest stalactite – 9 cm.

Figure 46: (right Chrysocolla and iron hydroxides coating mine timbers at a chute leading into a sulfide stope above.

7 level, Southwest Mine, vertical view - 1.30 meters.

Chlorides:

The exposure of unstable, non-sulfide minerals to the atmosphere during mining also resulted in the formation of post-mining species, often quickly. Nantokite, CuCl, and tolbachite, $Cu^{2+}Cl_2$ both copper chlorides, would react with the humid mine air to form paratacamite, $Cu_2^{2+}(OH)_3Cl$, in a matter of a few hours. In one case, a seven-centimeter cuprite nodule with a nantokite core developed a soft ram's horn of scaly paratacamite three centimeters high in less than 20 hours.

Figure 47: Paratacamite as a 8 cm., mushy extrusion from a cuprite nodule, which formed in less than 20 hours upon exposure to the humid mine atmosphere. In situ photo. 14 stope, 5 level, Southwest Mine

Copper:

As noted above, low pH, copper laden waters were common in the mines at Bisbee. In reality, these waters were solutions resulting from the oxidation of the variable and complex sulfide minerals and contained a variety of dissolved metals but, mostly iron and copper. The pH of these waters was quite variable, depending on the source and, of course, dilution, but pH ranges from below two and less than four were common. If not controlled, "copper water," (a local term that will be collectively applied herein) was bad stuff, yet it was both a bane and a blessing.

From a safety standpoint, there was nothing more uncomfortable than copper water in the eyes. From personal experience, we can say it was as painful as most any minor injury and temporarily debilitating. Skin burns from strong copper water or prolonged contact with it were unfortunately too common and took time to heal. These were much like burns from sulfuric acid.

Operationally, everything this water contacted was damaged, iron and steel most of all. Thus, care was always taken to control the flow of copper water to keep it away from the mine rail and pipe. The water was then channeled or pumped to recovery plants to extract the copper via iron replacement precipitation. Many millions of pounds of copper were produced at Bisbee in this manner.

However, once mining in an area was completed, the control of such water was no longer a priority and small areas would pond, often covering the now abandoned rail and all else left behind. Copper quickly began to replace the iron and steel. Initially, the original form of the replaced item is often apparent however, copper growth would continue as a direct precipitation of small crystals from concentrated solutions, now expanding the size of the copper mass and eventually obliteration the original form.

Figure 48: Front and back views of mine rail replaced by copper with the copper expanding well beyond the replaced steel, 2200 level, Campbell Mine, specimen – 10 cm.

Mass of copper thus formed could easily exceed by ten times the volume of the original iron. Oft times, miners would collect these copper masses and sell them as "native copper". In general, these pieces were relatively bright in appearance, reflecting their time in an acid environment. The expansive Junction and Campbell mines were the sources of most of these coppers, but all of the sulfide containing mines produced similar material.

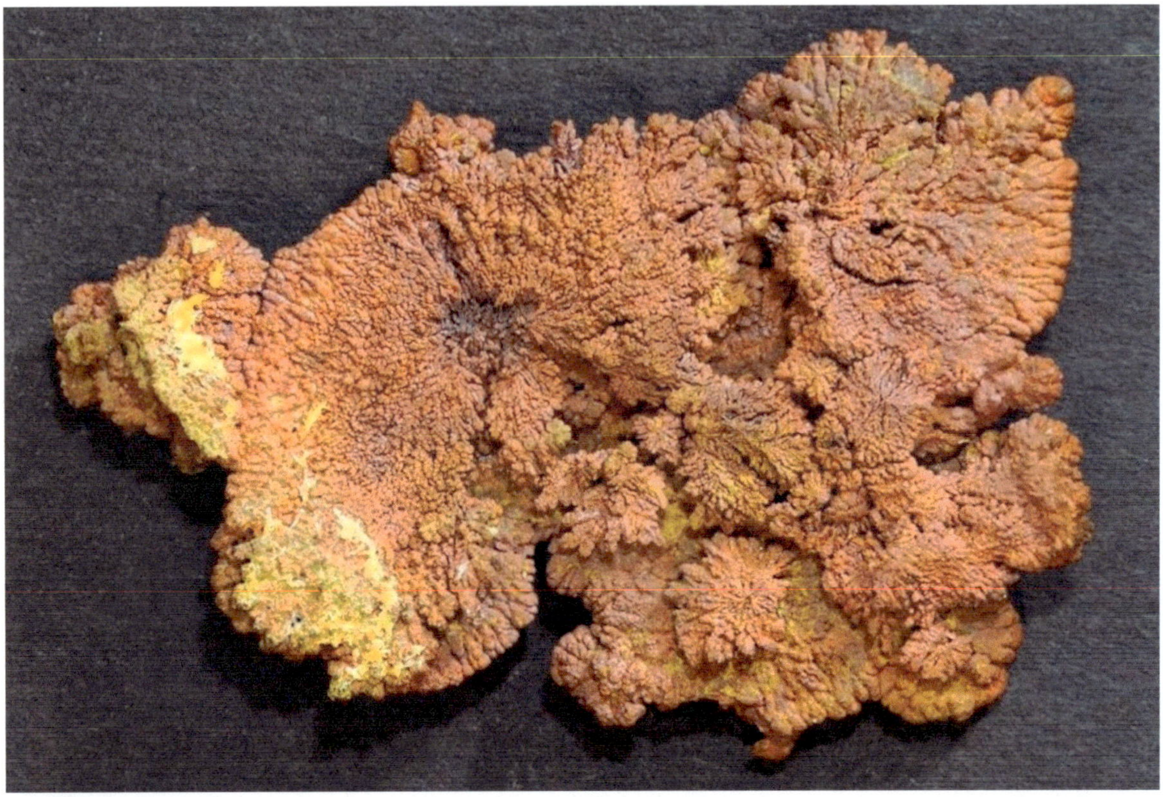

Figure 49: Copper which formed as a direct precipitation from solutions and not replacing steel or iron, 2200 level, Campbell Mine, specimen - 14 cm.

Figure 50: Gypsum on post-mining copper that precipitated directly from high copper solutions. 2200 level, Junction Mine, specimen – 9 cm.

In a few cases, copper would precipitate directly from concentrated solutions without forming around a replaced iron or steel core. Examples are shown above in Figures 49 and 50. These pieces tended to be composed of aggregates of tiny crystals that were more open, even spongelike or in radiating forms and could easily be confused with naturally occurring copper. This potential confusion was often compounded by one to two millimeter cuprite crystals that commonly accompanied these interesting pieces.

Copper directly precipitated on mine timber was found in the Lavender Pit Mine as it encroached upon old underground working of the Sacramento Mine. To preclude the unpleasant task of pumping these acidic waters from the pit, mine planning efforts always kept a low point in the pit over abandoned underground workings, which served as a drain for the mine. All of this water was collected on the 1600 level of the Sacramento Mine and directed to a pump station on the 1800 level Junction Mine for recovery. Over time, copper to five millimeters thick was deposited in voids of the usually badly broken timber. Much of this material was collected by the mine workers as the pit advanced.

Figure 51: Copper deposited on mine timber. Lavender Pit Mine; longest piece – 11 cm.

Fire developed minerals:

Mine fires were a constant problem in the district once the primary sulfides were exploited to any degree (Mitke, 1920). Pyrite would generate substantial amounts of heat during decomposition to the point of occasionally spontaneously combusting and igniting the supporting timber. The burning sulfides were a real problem because with adequate oxygen, they could burn for a very long time, emitting toxic gases, while denying access to ore in the fire area. Many of these fires defied all efforts to extinguish them and burned for years. Mine fires in both the Briggs and Campbell were contained, but still burning at the time the mines were closed.

Minerals produced as a result of mine fires at Bisbee have not been thoroughly studied. The few species recognized have been noted more by chance than by design. This is because of the lack of access to the areas under conditions that allowed study. Unlike Jerome, Arizona, the fires in the mines at Bisbee were never exposed to the surface by mining. While a number of fire areas were later mined underground, they had been flooded up to the time of reentry or had been filled with a slurry of mill tailings to allow access (Sherman, 1918; Mills, 1958). This destroyed the soluble minerals that might have formed. However, anthonyite from the Dallas Mine may well have been formed as a result of the sulfides that burned in a nearby stope during a mine fire (Anthony, et al. 1995).

The 1948 fire in the Campbell mine, which started on the 2200 level, spread quickly and soon reached the 1300 level. It was eventually vented to the surface through the Oakland shaft, which had been developed for that reason (Mills, 1958) and much of the fire areas reclaimed. Portions of the 1300 level were still very hot, if not burning, in the mid-1970s. The workings closest to the fire, through which the hot gases passed, had their limestone walls somewhat calcined, producing a very impure lime (<30% available CaO on analysis). Post-mining gypsum in a nearby area was converted to anhydrite, and in another area, tiny sulfur crystals were deposited on the mine walls. No other fire related minerals were noted during the single visit by one of the authors in 1975 to the accessible portions of the exhaust workings. Dangerously high levels of carbon monoxide gas and temperatures in excess of 55° C prevented a complete inspection.

SUMMARY:

The post-mining mineralogy of Bisbee is incomplete, at best. During the operational period of these wonderful mines, few took the time to look at was so common around them, rather the focus of the miners and collectors was always on the "Bisbee Classics." Today, some 40 + years after the mines closed, most sulfide zones are deep underwater and the rest of the mine workings are far too dangerous to even consider entering, thus precluding the recovery of new material . Perhaps the Lavender Pit may produce additional, if not new species, but it too is hazardous because of unstable and raveling pit walls.

Yet, in spite of their general unattractiveness and typical instability, some specimens were collected during operations. Perhaps upon further examination these will add to the information base in this neglected area of Bisbee's mineralogy.

BIBLIOGRAPHY:

ANTHONY, J. W., WILLIAMS, S. A., BIDEAUX, R. A., and GRANT, R. W. (1995) Mineralogy of Arizona, 3rd edition. University of Arizona Press, Tucson, Arizona, 508 p.

BATEMAN, M. N. and MURDOCH, J., (1914) Secondary enrichment investigations; notes on Bisbee, Arizona. Unpublished notes, Harvard University Mineralogical Museum files, 240 p.

BRYANT, D. G., and METZ, H. E., (1966) Geology and ore deposits of the Warren mining district; in *Geology of the Porphyry Copper Deposits, Southwestern North America,* edited by S. Titley and C. Hicks. The University of Arizona Press, Tucson, Arizona, 189-204.

DOUGLAS, J. (1909) Conservation of natural resources. Engineering and Mining Journal, 87, 1202.

GRAEME, R. W., (1981) Famous mineral localities: Bisbee, Arizona. *Mineralogical Record,* **12,** 258-319.

GRAEME, R. W., (1993) Bisbee revisited, an update on the mineralogy of this famous locality. *Mineralogical Record,* **24,** 421-436.

GRAEME, R. W. III, GRAEME, R. W. IV, GRAEME, D. L. (2015) an update on the minerals of Bisbee Cochise County Arizona. *Mineralogical Record,* **46,** 627-641.

MERWIN, H. E. and POSNJAK, E. (1937) Sulfate encrustation in the Copper Queen mine, Bisbee, Arizona. American Mineralogist, 22, 567-571.

MILLS, C. E. (1958) *Notations from annual reports (Copper Queen Consolidated Mining Company, Phelps Dodge & Company and Phelps Dodge Corporation) years 1909-1950.* Unpublished, Phelps Dodge Corp. files, 72 p.

MITCHELL, G. J. (1921a) Rate of formation of copper-sulfate stalactites. A.I.M.E. Transactions, 56, 64.MITCHELL, G. J. (1921b) Rate of formation of copper sulfate stalactites. Mining and Metallurgy, 170, 33.

MITKE, C. A. (1920) A history of mine fires in the South-West, part 1. Mining and Scientific Press, 121, 155-160.

SHERMAN, G. (1918b) Methods for controlling fires at the Copper Queen mine. A.I.M.E. Transactions, 59, 318-325

TENNEY, J. B. (1913) Unpublished report on 2,200 hand specimen and thin section determinations. Phelps Dodge Corp. files, 2,200 p.

Post-mining aragonite coating a one-gallon bucket on the 6 level of the Southwest Mine

Made in United States
Cleveland, OH
05 November 2026

47664967R00024